EarthCards

Postcards You Can Sign and Send to Save the Earth

Write for Action Group

Conari Press
Berkeley, CA

Important Note

Like so many of our culture's products, planned obsolescence is built right into this book. If you use it, by its very nature it will become at least partially outdated as corporate and political interactions with the environment improve. We think that the satisfaction we get from seeing the issues resolve and drop off the list is well worth the hassle of continually updating the book!

In any event, we'll do our best to stay current, and to keep our facts and figures accurate. But things in this field change quickly, so if you see in today's paper that President Bush has joined Greenpeace, and McDonald's has served their first McVeggieBurger–just smile, and keep those *EarthCards* for souvenirs!

Table of Contents

Preface

Average Americans like you and me often don't realize how much clout we really have. For some reason we assume that someone else–the imaginary "they" rather than us, or friends or our neighbors–have the power to change the world. "After all," we each wonder, "why should anybody listen to me?"

The answer is simple: it's our world too. We have a right to protect it for ourselves and our children.

Unfortunately, as more of us become convinced that we're too "small" to make a difference, fewer of us bother raising our voices. We don't even take the time to write letters. The regrettable result is that influential business leaders and politicians never find our what's *really* important to us. Sure they read the polls. But they've learned to be cynical. Anyone can answer a pollster's question effortlessly; only true believers will invest in a stamp and a trip to the mailbox.

But many congresspeople have so much respect for folks who send letters that they consider the opinions expressed in each one to be worth 100 votes! That's power.

So have you written your representative lately?

Have you written to auto makers about the pollution generated by cars?

Have you told the President what you think about his environmental stance?

Have you let the CEO of Procter & Gamble know you want a phosphate-free laundry detergent?

With *EarthCards* you can express your opinion quickly and effectively.

This well-written package is designed to give you maximum impact with a minimum of effort. You don't have to research all the issues in depth, to write a whole letter or look up any addresses. You just sign your name and send the cards.

Use your power. Let people in business and government hear your voice. If we all speak up, you can be sure they'll listen.

-John Javna,
The EarthWorks Group

Foreword

Hi! We're Ben & Jerry! Although we're just a couple of guys who make ice cream for a living, we like the environment as well as the next guy (as long as the next guy isn't James Watt). And that's why we're writing a foreword for this book–because we think that it might actually make a difference.

Also, we tend to like ways of getting things done that are unusual or innovative. And this book is both. Who ever heard of a book of pre-written environmental postcards? But it's a great idea, and if lots of us use it, it will work! At our company, if we get as few as 15 or 20 pieces of mail on a given subject, that topic becomes a priority for us to deal with. And we think that lots of companies feel that way (although most politicians probably require a bit more mail than that to get 'em jump-started).

We like postcards, too. One particular card has had a big impact on us. A few years ago we heard from a fan of the Grateful Dead band, suggesting that we create a Cherry Garcia ice cream. We did just that, and now it's one of our most popular flavors! Not bad results, for a single postcard!

Each of these *EarthCards* alone may not have quite as much effect as our Cherry Garcia card had on us. But together, they can make all the difference in the world, and for the world!

-Ben & Jerry
Vermont

All About *EarthCards*

Just What Exactly Is EarthCards?

The *EarthCards* concept is simple. We realized that although personal actions such as recycling newspapers, saving water and driving an energy efficient car, are vital, healing the earth will require a *political* and *corporate* commitment as well. We believe that by sending *EarthCards*– lots of *EarthCards*–to carefully selected recipients, we can help decision-makers in business and government make the choices that will slow, then stop and even reverse the ecological destruction of our nation and our planet.

Your *EarthCards* book is a tool that makes it easy for large numbers of concerned citizens like ourselves to voice our concerns on a variety of pressing environmental issues. And variety there certainly is; some-times the sheer number of problems facing us is boggling, which may be why so many of us mean to sit down and write those letters, but rarely get around to doing so. And even if we decide which of the hundreds of equally critical letters to write, we then have to determine whom to send them to and what to say.

But *EarthCards*, in effect, does our homework for us. By choosing the issues and the recipients, *EarthCards* will deliver a concentrated out-pouring of mail to a limited number of influential individuals (narrowing them down wasn't easy, as our "Hard Choices" section describes).

When each of us writes an occasional letter on the issue of our choice to a politician or executive, the effect is rather like the watering of a lawn sprinkler. But by using *EarthCards*, we can direct a barrage of cards as powerful as the stream of a fire hose. And by giving the people in power a piece of our minds, we can change theirs!

Who Is The Write For Action Group?

If you're reading these words, you're a member of the Write For Action Group. Everyone who has helped to create *EarthCards*, and everyone who uses it, is an important part of this innovative strategy for making the planet a healthier place for all of its inhabitants.

The Write For Action Group was designed along the lines of the Earthworks Group, the people who wrote the popular book *50 Simple Things You Can Do To Save The Earth*. As with the Earthworks Group, a few imaginative minds came up with the *EarthCards* concept, then formed the Write For Action Group by contacting and collaborating with a variety of knowledgeable environmental organizations and individu-als.

Some of the incredibly helpful folk that we'd like to thank for their help in selecting and writing *EarthCards* include:

Audubon Society; Ben & Jerry's Ice Cream, Ben Cohen, Jerry Greenfield; Center for Science in the Public Interest, Beth Kaufman; Citizens Acid Rain Monitoring Network; Citizens Clearinghouse for Hazardous Waste, Lois Gibbs; Clean Water Action Project, Ken Brown; The Cousteau Society, Clark Lee Merriam; David Harp; Earth Island Institute, John Knox; Earthworks Group, John Javna, Julienne Bennett; Environmental Action Coalition, Nancy Wolf, Tim Forker; Environmental Defense Fund, Joel Plagenz; Garbage Magazine, Janet Marinelli; Greenhouse Crisis Foundation; Greenpeace, Campbell Plowden and many others; Institute for Alternative Agriculture; National Toxics Campaign, Gary Cohen; National Wildlife Federation, Stewart Hudson; The Natural Resources Defense Council, especially Chris Calwell; Nature Conservancy, Bob Klein, Carol Baudler; New York Rainforest Alliance, David Katz, Ivan Ussach; Rainforest Action Network, Pam Wellner, Annie Szvetecz; Rocky Mountain Institute, Hunter Lovins; The Sierra Club, Rich Hayes; Smithsonian Institute, Katy Moran; Turner Broadcasting System, Barbara Pyle; 20/20 Vision National Project, especially Jeremy Sherman; The Wilderness Society, Mary Hanley; World Bank Information Center, Chad Dodson; World Resources Institute, Shirly Geer; World Wildlife Fund, Bill Eichbaum, Jim Leape; Worldwatch Institute.

The EarthCards Format

We present brief background summaries entitled "The Problem" and "The Solution" for each of nine general topics: Global Issues; Rainforest Preservation; Toxic Waste; Recycling; Consumer Products We'd Like to See; Media Attention to the Environmental Crisis; Wilderness Preservation; Energy Efficiency; and Corporate Commitment and Social Responsibility. Then we briefly describe the *EarthCard* or *EarthCards* that relate to that topic. Some are followed by suggested P.S.s you can add and Write for Action Resources for additional action.

What EarthCards Isn't

We've tried to keep the scope of this project focused in order to keep the book short and the price down (recycled perforated card stock is plenty expensive, even without loads of additional pages). So *EarthCards* is not a book full of fascinating environmental facts and figures, or suggestions for personal action. If that's what you want, read *50 Simple*

Things You Can Do To Save The Earth ($4.95, available at bookstores, or send $5.95 [Ca. residents send $6.31] to: Earthworks Press, 1400 Shattuck Ave. #25, Berkeley 94709).

Nor does *EarthCards* provide a credo of deep ecology, or a profound general explanation of the environmental crisis. For that, we recommend Barry Commoner's excellent *Making Peace With The Planet* (from Pantheon Books, and available at better bookstores everywhere). And we just didn't have room to include information on how to use our consumer dollars in the most ecologically-aware manner–Penguin Books' *The Green Consumer* did that.

No, *EarthCards* is a book of sign 'em, stamp 'em, and send 'em postcards. So get ready to send a card to your Mother–Mother Earth!

How To Use EarthCards

As Ben & Jerry and John Javna noted in the Foreword and Preface, if a company gets even a small amount of mail on a particular subject, they usually take notice. If they get a large amount, they just may take action. A great example of this is noted on the back cover: after receiving fewer than 8,000 letters, Starkist Tuna decided to reverse a dolphin-damaging policy of many years' duration. So *EarthCards* will work–we're sure of that–if you use it! Here's how:

Flip through the list of topics in the Table of Contents. Pick the topic or topics that feels most important to you right now. Turn to the appropriate page to read the brief summary of the issue and the short discussion of each *EarthCard*. Then turn to the perforated postcard section. Read the *EarthCards* on your chosen topic, (find them by matching the graphics from the topic section) tear them out, sign them (with a PS if you like), stamp them, and send them! You've just made a difference!

How To Send EarthCards to Congress

Letters and cards to politicians are most effective when you are one of their constituents; that is, when they have to worry about whether or not you'll vote for them. So some of your *EarthCards* will go to your very own two Senators in Washington, and to your one U.S. Representative. Those of us who don't follow the political scene too closely may need to find out their names before sending these *EarthCards* (don't worry about the addresses–they're there already).

Fortunately, a single local phone call will locate these names for you. Simply call either the town or county clerk listed in the front of your local phone book, and request the names of your U.S. (not state) Senators and Representative. You can also call any of your local state Senators and state Representatives, all of whom will be listed in the front of the phone book. In fact, many phone books list the names of Congresspersons too. As a last resort, call your local library–librarians can tell you in a flash!

Once you've obtained their names, why not write them in (just after the words "The Honorable [first name] [last name]") on the appropriate EarthCards right now.

Hard Choices

Before you start signing, stamping, and sending, you may want to know a bit more about how we chose what to include in *EarthCards*. We began by asking a wide variety of environmentally aware people what they thought were the most pressing topics facing us all. In addition, we studied the many recent newspaper polls on ecological issues (from *USA Today* to the Associated Press to the *Wall Street Journal*). Between these two sources, we came up with our list of nine general issues.

With these nine areas in mind, we started to contact both specific-subject groups (like the Citizen's Acid Rain Monitoring Network or the Rainforest Action Network) and more general ones (like the Sierra Club or the Natural Resources Defense Council). They helped us to choose and shape each *EarthCard*.

But it still wasn't easy. Sometimes, we'd zero in on what seemed like a great idea for an *EarthCard*, only to find a hidden flaw or controversy. The proposed National Bottle Bill, for instance, looked good until we spoke to groups that specialized in recycling issues. Then we learned that the bill would, by placing a return deposit on aluminum cans, eliminate the most profitable component of local recycling centers. Without aluminum cans (which under a bottle bill would be returned to the stores that sold them), recycling centers would shut their doors.

Eventually, we came up with the final selection. Most *EarthCards* relate to consumer issues, a few to political legislation. We tried to choose issues and target *EarthCard* recipients in a way that would maximize effectiveness. Many factors had to be considered, from the symbolic importance of an issue to the length of time it might take to be resolved. Each individual and organization involved in the project had a personal "wish list" of *EarthCards* and many compromises had to be made.

We realize that many fine opportunities for *EarthCards* have been omitted. However, with your help and enthusiastic response, there is something we can do about that. We plan to come out with a series of sequels, with each one focusing on one of our general topics. So keep your eyes peeled for a book of Rainforest *EarthCards*, a Toxic Waste *EarthCards* book, and so on!

Some Questions We'd Like To Answer

Are Printed Postcards As Effective As Hand-Written Letters?

No. Studies of lobbying and letter-writing campaign techniques indicate that a mass-produced postcard has only ten or 20% as much impact as a hand-written letter in an envelope. But we're convinced that the sheer volume of mail that together we can generate with *EarthCards* will far more than make up for the decreased impact of each card. And there are ways to make each *EarthCard* more effective, as we discuss below. Of course, if reading an *EarthCard* really fires you up on a given issue, please feel very free to either re-write our card in the form of a hand-written letter, or use it as an outline for a longer letter.

How Can I Make EarthCards More Effective?

Our research indicates that your *EarthCard* will have more of an impact on its recipient (especially when that person is a politician), if you add a brief hand-written P.S. For some *EarthCards*, we've suggested a P.S. And here are some general P.S.'s that could be added to any *EarthCard*. Please make up any variations of your own that feel right.

P.S. This is really important to me, as a (father/mother/grandparent).

P.S. I truly believe that this is a matter of life and death.

P.S. I await your response and have begun to organize my friends and neighbors to broaden support for this issue.

Return Address: And make sure to include your name and address in the space provided–it adds further credibility to your *EarthCard*.

What If I Don't Think You've Gone Far Enough?

Most environmentally-concerned Americans should be willing to support the positions we take in the *EarthCards*. Some of you may want to go further. A P.S. like one of the following may be just what is needed to strengthen the point.

P.S. I think that all non-rechargeable dry batteries should be phased out by the year 2000.

P.S. I'd even support banning fossil fuel cars entirely in areas that don't meet Clean Air Act standards.

P.S. I'd like to see CFCs banned from the marketplace by 1992.

About The Nine Topics

In some ways, our general topics are not as clearly separate as they sound, since ecological catastrophes tend to have little respect for nice, neat, labels. The toxic waste crisis, for instance, can be alleviated through both recycling efforts and creation of more ecologically-sensitive consumer products. An increased corporate commitment to the environment will reduce toxic waste, increase recycling, conserve energy, deliver the consumer products that we'd like to see, and help to resolve global issues like the greenhouse effect and thinning of the ozone layer. And some of our recycling *EarthCards* (especially the one to McDonald's) could almost have fallen into the "Consumer Products We'd Like To See" category, and vice-versa.

10

Certain topics are included implicitly. Acid rain is not one of our topics, but many *EarthCards* (like the toxic waste cards or the ones to car makers) make suggestions that would have a beneficial effect on the acid rain problem.

To help you find specific *EarthCards*, each one will bear the logo of the general topic that most closely relates to that card. But rather than thinking of these topics as mutually exclusive labels, please consider them as a convenient "eco-graphic" device by which to group and find the *EarthCards*.

Four Crucial Points for the Planet

There are four general points that no book on environmental issues can afford to miss. Briefly, they are:

• We need to prevent pollution and ecosystem destruction before it occurs, rather than just trying to clean it up after it's happened.

• We must increase our ability (and our desire) to use energy more efficiently, rather than increasing our exploitation of the Earth's limited energy resources.

• We have to end our denial of the hidden costs of our present lifestyle. These hidden costs include the pollution by our use of electricity and fossil fuel, and the price of cleaning up our toxic wastes and recycling our refuse.

• And finally, it's not pleasant, but it's true: cleaning up the planet is not going to be cheap. If you hadn't cleaned, painted, or repaired your own house in 30 or 40 years, you probably wouldn't be surprised to find that you had to spend some time and money getting it back into livable shape. Planets are not that different from houses, and unfortunately we haven't been doing our housecleaning since the Industrial Revolution began pouring out pollution, centuries ago.

So it's time to hire whatever help we need, and take the time for a good spring cleaning to welcome in the millennium. Since our planetary budget for weaponry has been averaging a hefty trillion dollars a year, a few hundred billion a year for the environment shouldn't phase us, since a cleaner planet is a healthier planet, and a healthier planet is a happier planet (with a reduced need for arsenals).

A number of our discussions in the general topics and *EarthCards* will implicitly or overtly reiterate these four points. For a few more comments on the need for prevention rather than clean-up of pollutants, please read the Toxic Waste section (page 16). For an expanded explanation of the concept of hidden costs, please see the Recycling section (page 17). Want more on energy efficiency versus exploitation? See the Wilderness Preservation and Energy Efficiency sections on pages 24 and 26. And if you're willing to think about paying the price, read the Corporate Commitment section on page 28.

Global Issues

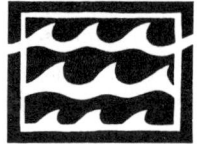

The Problems: These are issues that involve the entire planet. Most notably: the *greenhouse effect* (whereby increases in the planet's average temperature may impair agriculture worldwide, and destroy the earth's coastal regions when the polar icecaps begin to melt), and the *decrease in the ozone layer* (which will not only raise the rate of skin cancer, but may disrupt the entire food chain by killing off creatures unfortunate enough not to have access to sunscreen and baseball caps).

The Solutions: There are already solutions to both of these problems. Unfortunately, the U.S. government has not been a particularly enthusiastic supporter of them, even though our 5% of the planet's population accounts for an estimated 20% of the greenhouse effect. Other major industrial powers (with the glaring exception of the Soviet Union) have already committed to reducing the levels of "greenhouse gases" (largely caused by the burning of fossil fuels in factories, power plants, and your car) and CFCs (chlorofluorocarbons, the major eaters of ozone). We need to join them, right away, instead of procrastinating with talk of "further research."

The EarthCards

President Bush: Our Earthcard reminds him of his promise to be the "Environmental President," and stresses the urgency of the need for immediate action on the issues of global warming and the thinning of the ozone layer.

Suggested P.S.: *Please listen to what William Reilly has to say (rather than Sununu) and act on it!*

Global Warming: Global Warming Bill H.R. 1078, sponsored by U.S. Representative Claudine Schneider (R-Rhode Island), is probably the best legislation on this topic. It has the bi-partisan support of 144 members of the House, and has been referred to 12 different congressional committees, which is both a sign of its comprehensiveness and a forewarning of a long legislative battle. Most pro-environmental statutes (like the Clean Air or Clean Water Acts) pass into law only after protracted haggling, so even if H.R. 1078 is unlikely to be enacted promptly, it's a decent start on a complex and important issue.

One *EarthCard* in support of this bill goes to your U.S. Representative, and one similar card to each of your two Senators. The Senator cards will suggest that a thorough study of the contents of H.R. 1078 would help create a single, more comprehensive bill combining Senator Gore's S. 201 and Senator Wirth's Global Warming Bill S. 324 (which places too much emphasis on nuclear power, a costly and unnecessary part of the legislation).

Suggested P.S.: Appropriate for all of the above cards: *Let's lead the rest of the world on these issues, not follow it!*

Write for Action Resources: Greenhouse Crisis Foundation, 1130 17th St, NW, Suite 630, Washington, DC 20036 (202) 466-2823.

One More Global Issue

The Problem: This one isn't strictly on global warming or the ozone layer, but it's still clearly an issue of worldwide scope. The World Bank, owned by 150 nations, funds some of the largest projects that the earth has ever known (usually in very poor countries, and often relating to energy production). So what's the problem? Well, the World Bank has not always been very environmentally conscious in their choice of projects. It has just begun to factor ecological considerations into project analyses. Plus, many small countries have ended up owing large sums of money to the Bank for energy projects that do more ecological harm than good in the long run.

The Solution: The members of the Bank know what they need to do, but they need lots of support doing it. The Bank needs to spend at least as much money helping developing nations save energy as it spends helping them generate energy. Also, since so many countries owe it money, it is in a perfect position to conduct what are called "Debt-for-Nature Swaps" in which the Bank trades forgiveness for some of the debt owed it. In exchange, countries promise to protect by law ecologically sensitive parcels of land, such as rainforests or wildlife refuges.

The EarthCard: This card will be sent to Mr. Barber Conable, President of the World Bank, requesting that the Bank enlarge its environmental committee and prioritize long-term environmental issues when considering projects. It also suggests that the World Bank implement large-scale "Debt-For-Nature" swaps with its debtors, if environmentally appropriate.

Write for Action Resources: For more information, contact the World Bank Information Center, 2000 P St. NW #515, Washington, DC 20036 (202) 822-6630.

Rainforest Preservation

The Problem: If the planet were a person, the rainforests would be her lungs. But the remaining rainforests of the world (40% have been destroyed by humankind in the last 100 years) don't simply convert unbreathable carbon dioxide into life-supporting oxygen, thus helping to counter the greenhouse effect. They also provide a unique habitat for literally half of the world's plant and animal species, some of which may well be medically or agriculturally useful, and all of which have a perfect right to survive on the planet. From Brazil (the best-known rainforest) to Burma, rainforests continues to be burnt down at a terrifying rate.

The Solution: Destruction of the rainforests of the world must be opposed wherever it occurs. This means that a price must be paid. For example, Burma, Malaysian, and Thailand see their rainforest wood as a valuable and exploitable natural resource. The U.S. members of the International Hardwood Products Association are just trying to make a profit by importing and selling hardwood that we all use. Who will pay the various prices that preservation requires? Can we somehow share the cost? What is fair?

These are tough questions. We can't answer them here, but we'd also be remiss to ignore them. As Write For Action Group member David Harp points out in his book *MetaPhysical Fitness:* "We must save the whales, and the whalers...." While we all grapple with the issue of responsibility, there are some specific points of intervention that can be attempted.

The EarthCards

The Ecuadorian Rainforest: The giant U.S. oil company Conoco wants to build a road and oil pipeline into Ecuador's Yasuni National Park and the Waorani Indian Nation, both located in a relatively unspoiled spot in the heart of the Amazonian Rainforest. As Stewart Hudson of the National Wildlife Federation points out, although Conoco explicitly intends to try to prevent damage to the native culture and the eco-system, the fact remains that whenever a road has been built into the rainforest, rampant and uncontrollable colonization (often by desperate, poverty-stricken peasants) follows. Conoco wants to be seen as being environmentally-conscious; that's probably one reason why they committed to double-hulling their oil tankers after the Exxon Valdez disaster. Let's help them make the right choice on this issue!

The Burmese and Malaysian Rainforests: Although less often publicized than the Amazon region, the rainforests in Burma and Malaysia are also in the process of being clear cut, with the usual attendant human rights abuses of the local indigenous peoples. In fact, many European countries and Australia have already called for a boycott of hardwoods from this area, unless the seller can prove that the wood was obtained by sustainable forestry practices. Most of the rainforest action groups in this country agree that the Virginia-based International Hardwood Products Association should be urged to stop buying hardwoods from this part of the world until the ecological and human rights issues have been better addressed (this would include wood from Thailand also, since Thailand acts as middleman for much of Burma's wood sales).

Write for Action Resources: Rainforest Alliance, 270 Lafayette St. #512, NY, NY 10012 (212) 941-1900; the Rainforest Action Network, 301 Broadway, San Francisco, CA 94114 3 (415) 398-4404, *The Rainforest Book* by Scott Lewis, (available by sending $7.95 to: NRDC, Box 1400, Church Hill, MD 21690).

Toxic Waste

The Problem: According to a recent *USA Today* poll, 67% of Americans characterize themselves as "very worried" about hazardous waste. And it's no surprise that more folks were very worried about toxic waste than any other single environmental issue, since 4.57 billion pounds of chemicals were poured into our nation's water, air, and ground in the year 1988 (the most recent year for which figures are available)–nearly 20 pounds per citizen!

Despite concern, the toxic waste situation hasn't improved since the first Earth Day back in 1970. Levels of only a few major pollutants, such as lead, PCB, and DDT, have decreased dramatically–and these are the pollutants that have been banned outright, not the ones that have been "controlled." Attempting to control the effects of pollutants (whether through smokestack scrubbers to reduce the toxic wastes that cause acid rain, or catalytic converters that reduce auto emissions) is far less effective in the long run than changing the processes that cause the pollution in the first place.

The Solution: The Resource Conservation and Recovery Act (RCRA, pronounced "Rick-ra") is the set of laws that regulates hazardous and solid waste disposal. It will be coming up for re-authorization within the next year or so, and could be strengthened to improve the current toxic waste situation. In order for this to happen, RCRA must emphasize reduction of hazardous waste before it occurs, rather than disposal afterwards. The existing version of RCRA will concentrate on building toxic waste incinerators, which many scientists believe will simply spread the poisons into the atmosphere, and on forcing states to create sites for hazardous waste dumping (whether they want to, or not).

But there is still time to improve RCRA, if we let our politicians know how strongly we feel. A RCRA based on reduction also will be good for the economy, because it will force companies to face the hidden costs of waste disposal that the government is now covering with our taxes.

The EarthCards: One will go to our U.S. Representative, and one each to our Senators, stressing the importance of re-formulating RCRA as a waste reduction bill, not as a forced dump/incinerator placement bill.

Suggested P.S.: *I truly believe that this is a matter of life and death.*
Write for Action Resources: National Toxics Campaign, 37 Temple Place, Boston, MA 02111 (617) 482-1477; Greenpeace USA, Toxic Campaign, 1436 U Street NW, Washington, DC 20009 (202) 462-1177.

Recycling

The Problem: Recent polls show that between 80% and 90% of Americans support the idea of recycling, and curbside programs and drop-off recycling sites are springing up around the nation. They can be very effective, but in order to really work, two elements must be addressed: hidden costs of recycling (or disposal), and the truly cyclic nature of recycling.

Use of many of the manufactured products in daily life (especially the petrochemical ones) can be likened to borrowing money from a loan shark. Loan shark money is easy to get, and convenient in the short run, but there is a major cost in the long run, which becomes abundantly obvious when "Rocky" shows up to break your fingers. Likewise, many of the manufactured items we use daily have a hidden cost in health dangers and disposal problems, a cost that we rarely see because it is not calculated into the point-of-purchase price of the product.

The Solution: Recycling can reduce the costs of disposal, because the re-use of existing materials is often more cost-effective than the creation of new items (recycling an aluminum can, for instance, saves 95% of the energy necessary to make a new one from scratch). But recycling has mechanical and organizational costs of its own, and these should (in the best of all possible worlds) be calculated into the initial cost of the item, as indeed *all* hidden costs should be.

The determination of all costs is called "cradle-to-grave analysis," because it includes the usually hidden costs of both production and ultimate disposition. We need our corporations and government to make a cradle-to-grave analysis part of every decision-making process. This will mean that the people who use a product will end up paying more when they buy it instead of taxpayers having to pay for the eventual cleanup of the product's uncontrolled disposal later on. And of course this would eliminate taxpayers' need to pay the health costs that usually result from said uncontrolled disposal, as in the Love Canal or other toxic waste sites. Finally, this more honest initial price will make the use of hazardous products less attractive in the first place, so it will *prevent* environmental damage. And we'd much rather prevent it than clean it up later on.

Putting The "Cycle" Into Recycling: When many of us think of recycling, we think of separating our newspapers from other trash, and dropping them at the curb or local recycling site. But that's only the first part of the cycle. Once the papers have been collected, ground into pulp, de-inked, and made into new (recycled) paper, somebody has got to

17

buy that paper for the cycle to be complete. So to *really* be recyclers, not only should we recycle everything that can be, we have to take responsibility for creating a demand for products made from recycled materials.

A Note On Definitions: There are currently no standardized definitions for recycled products. What difference does that make? Well, in paper, for example, paper mills can take scraps and remnants of wood and paper from the production process of non-recycled paper, re-use it, and label their product recycled, even though the only waste product used in the process was pre-consumer waste (material that never left the paper mill). Unfortunately, only when manufacturers use post-consumer waste (that's the paper that you've already used once, then put in the recycling bin) does a market exist for your old newspapers! Therefore, when buying recycled products, it is important to know that post-consumer waste was used in the manufacturing process.

The EarthCards

On Batteries: Americans use a billion and a half disposable alkaline batteries every year, with each battery composed of an average of 1% mercury, an exceptionally hazardous waste that cannot be safely disposed of by land, sea, air, or fire (they are, of course, disposed of–but not safely–which represents a tremendous hidden health and environmental cost). Almost no disposable alkaline batteries are recycled in this country, although it is routinely done in Europe. In order to be cost-effective, large numbers of batteries must be available for recycling, which is not currently the case. Furthermore, technology exists (and is increasingly being used in Europe) to manufacture mercury-free batteries.

Our *EarthCards*, to Duracell and Eveready, two of the largest battery makers, express appreciation for their efforts to lower the mercury content of batteries, but also make two requests. We ask that they implement a disposable alkaline battery recycling program by offering to buy back used batteries which then, when available in sufficient numbers, can be economically recycled. Such a program could be designed similarly to a current one for recycling car batteries in which K-Mart pays customers to bring in their used car batteries which are then stored for collection. Battery companies could have K-Mart, or another national chain that sells their products, offer the same type of service for alkaline batteries. We also call upon them to use the European mercury-free battery technology in their manufacturing. Removing pollution from the process is a better solution than cleaning it up later!

On Books: This *EarthCard* goes to the CEO of publishing giant Bantam/ Doubleday/Dell, expressing our appreciation and consumer support for their recent interest in using recycled paper in some of their books, and suggesting the use of other low-impact publishing techniques such as soy ink and recyclable binding glue. By letting them know that we consider recycled paper in a book to be a real plus (even though we may have to lower our standards for brightness of paper), we can encourage them to switch more and more of their books to recycled paper. And if it works for them, other publishers will follow suit, greatly increasing the demand for recycled paper.

It is estimated that the use of paper with a 50% recycled content saves, per ton, 17 trees and 7,000 gallons of water. In addition, 60 pounds of airborne waste and 3 cubic yards of solid waste are avoided. Were Bantam/Doubleday/Dell to use this type of paper for even half of their books, they would save an estimated 250,000 trees and 100,000,000 gallons of water annually, and eliminate 850,000 pounds of air polluting effluent and 43,000 cubic yards of solid waste!

To McDonald's: McDonald's new McRecycle USA program is certainly a step in the right direction. Slated to cost 100 million dollars (out of McDonald's 17 billion dollar share of the fast food market), it will increase trash separation facilities to include 2000 restaurants (only 450 currently separate for recycling), test in 500 restaurants carryout bags made of 65% recycled newspaper and change its purchasing requirements for paper and cardboard goods that will make them more easily recyclable.

Given the size and visibility of McDonald's, these efforts are appreciated. Creating a demand for recycled products is crucial. But there are at least three other steps that are vital to making their environmental program a truly effective one.

In order to most effectively deal with waste disposal, McDonald's needs to reduce its total amount of packaging per food item sold, especially the use of polystyrene. It also need to reconsider its decision to begin installing incinerators (called Archie McPuffs) for styrofoam and other waste behind their restaurants, since surveys show that incineration of styrofoam (especially when mixed with other food and packaging materials) can release a variety of toxic chemicals.

Lastly, we're asking McDonald's to offer a low cost, healthy, non-meat burger, a McVeggieBurger, if you will. Doing this would have a tremendous real and symbolic environmental value–the production of a pound of veggieburger consumes far less of the planet's limited energy and resources than does the production of a pound of hamburger (up to 100 times less water, for example, depending on the kind of vegetables used). And if McDonald's came out with a veggieburger, its competitors would soon follow suit.

Suggested P.S.: If you can truthfully do so, you might add something like: *I often buy (batteries/books/fast food), and your action on these requests will have a big influence on my purchasing decisions in the future.*

Write for Action Resources: For a free information packet on recycling resources, call the Environmental Defense Fund at 1-800-CALL EDF (225-5333). Or call your local public works department at city hall, and ask them about local recycling efforts. *Garbage Magazine* covers a wide range of recycling issues, as well as being a good monthly read–call 800-274-9909 for subscription information. If you are interested in setting up a local recycling program, see the Generic *EarthCard* on page 30, and contact U.S. Public Interest Research Group (USPIRG) headquarters at 215 Pennsylvania Ave. SE, Washington, DC 20003 (202) 546-9707. They'll connect you with your local USPIRG office which can help you set up a neighbothood program. Also, look for *The Recycler's Handbook* by the EarthWorks Group, $4.95, available in bookstores.

Consumer Products
We'd Like To See

The Problem: As consumers, we've become used to the dubious convenience of petro-chemical packaging (made of oil-based chemical materials like plastic), rather than packages made of organic materials like wood and glass. This packaging is usually neither re-usable nor recyclable. The products inside the package are not always environmentally benign either, but both packaging and content can be controlled by us consumers–if we choose to do so. If we don't express our preferences, the various industries that provide us with products will continue to give us what's convenient and profitable for them, not what's good for us.

The Solution: By educating ourselves about consumer/environmental issues, and by letting the businesses that we buy from know what we want, we can get merchandise that's better for us, and for the planet.

The EarthCards

Proctor and Gamble: With production and distribution of 83 brands in 38 product categories, P&G is one of America's largest businesses. They already have a stated commitment to improving themselves environmentally, and an aim of reducing packaging by 25% in all product categories. However, according to a *Garbage Magazine* estimate, P&G products also accounts for an amazing 1% of the nation's trash! Their penetration of the marketplace could make environmentally benign products easily available to everyone–if they'd just make the products!

Our *EarthCard* to them encourages their movement towards ecological conscientiousness, and requests that all their plastic products be made of 100% recycled plastic (as are some of their new Spic and Span bottles). This will stimulate the market for recycled plastic, and thus encourage collection and recycling. We also suggest they consider a variety of eco-products: a non-phosphate laundry detergent powder to reduce water-polluting algae blooms caused by phosphates; a cotton-based diaper insert for use with reusable diaper covers to reduce the need for disposable diapers, none of which are currently very biodegradable; a line of vegetable-based instead of petrochemical-based cleansers for dishwashers, windows, woolens, floors, cars, and other uses; and a high post-consumer waste content (100%, if possible), unbleached and thus poisonous dioxin-free line of paper products including toilet paper, facial tissue, and paper towels.

The Farm Bureau: If we are what we eat, Americans are in trouble. And we know it, since a recent survey by the American Farm Bureau (the largest farm trade organization in the country) discovered that three out of four Americans think farmers use more pesticides than they should. One might think this poll would motivate the Farm Bureau to help its 3.3 million members to decrease use of pesticides. Doing that would be easier now than ever since "low impact" agricultural practices have become more efficient (due in part to increased understanding of integrated pest management techniques that combine use of natural enemies of pests with limited pesticide use) and more cost effective (since organic and unsprayed foods have become popular, in part due to fears about pesticide contamination).

But rather than offering us pesticide-free food, Farm Bureau President Dean Kleckner has announced plans to mobilize a public relations campaign, with the intent of "educating" consumers on the virtues of pesticides and the safety of U.S. agricultural products. We think they should change their approach.

Scott Paper Company: Our *EarthCard* asks Scott Paper, one of the nation's largest paper product manufacturers, to provide us consumers with a widely available line of high post-consumer waste content, unbleached paper products including toilet paper, facial tissue, paper towels, and picnic ware, as well as a cotton-based diaper insert for use with reusable diaper covers. We also request that they research and attempt to market a non-polystyrene foam cup that can be used for hot and cold liquids.

Of course we'd like to see all of the paper makers provide recycled products. We've focused on Scott because they are so large that their actions would make such products available in every grocery store. Also, they have an inherent interest in social responsibility, as evidenced by their generous charitable donation policy.

Write for Action Resources: Until P&G, Scott and other big manufacturers come through with the desired products, most of them can be obtained from the Seventh Generation catalog of "Products for a Healthy Planet," at (800) 441-2538. Many such products are also available from Shaklee–look for their local sales rep in your phone book.

Media Attention to the Environmental Crisis

The Problem: With the noticeable exception of programming around Earth Day, the three major TV networks tend to pay less attention to the environment than they do to the latest crime or movie merchandising tie-in. When a particularly gruesome disaster happens (like the Exxon Valdez oil spill), it becomes a brief media event, while ongoing catastrophes (like the worldwide destruction of the rainforest or global warming trend) are relatively ignored. Only Turner Broadcasting System (TBS) appears to have a serious commitment to environmental programming. Under the direction of United Nations Global 500 Laureate Barbara Pyle (Vice President for Environmental Programming), TBS will program two new weekly shows: "Captain Planet and the Planeteers" (a children's show with an environmentalist hero) and "Network Earth" (a news show on pressing ecological problems and their solutions).

The Solution: The major networks have repeatedly demonstrated that they are sensitive to viewer demands as expressed by mail. Let's make sure they know what we want!

The EarthCards: One to the presidents of CBS, NBC, and ABC requesting increased environmental coverage of ongoing as well as acute situations. We also inform them that sponsorship of such programs would tend to dispose us favorably to their advertisers.

Suggested P.S.: If you're a TV fan, you might say something like (if true): *I often watch (CBS/ABC/NBC), but why do I need to turn to cable to get news on environmental issues?*

Wilderness Preservation

The Problem: As the population of the world (and the concomitant need for additional resources and energy) expands, the unspoiled (or less spoiled) portions of the planet shrink. Yet there seems to be an almost inborn human need for nature and natural settings. Polls indicate that more than half of us are "very worried" about wilderness preservation, even in the face of more immediately threatening ecological problems. Yet two of the most pristine remaining regions of the globe, the Arctic and the Antarctic, need immediate protection in order not to be exploited for their potential (and perhaps limited) natural resources.

The Solution: It appeared as though Antarctica was facing serious danger of mining exploitation from the 20 nations (including the U.S.) who signed the Convention on the Regulation of Antarctic Mineral Resources Activities (CRAMRA) agreement in 1988. Recently, however, signatories France and Australia have decided to withdraw support from CRAMRA, and instead subscribe to the creation of an Antarctic international park, with mining prohibited. The U.S. government still supports CRAMRA, but the Senate has not ratified the agreement–and, with our help, may never do so. Senate Joint Resolution 206 and House Joint Resolution 418 would reinforce world efforts for an international park, which is exactly what the Antarctic should be.

Despite its unique and beautiful coastal plain eco-system, the 18,000,000 acres of the Arctic National Wildlife Refuge is not protected from possible oil, gas, and mineral exploitation. According to the World Resources Institute, the generally accepted estimate is that 4 billion barrels of oil might potentially be found in the refuge. This same amount of oil could easily be conserved by raising the fuel efficiency of the average car by only 1% each year for the next 25 years. We further develop the idea of energy efficiency versus exploitation of increasingly scarce and expensive resources in the next section.

On The Endangered Species Act: The preservation of a diversity of plant and animal life is a sub-category of wilderness preservation. There is something terrible and perverse about the death of an entire species–the loss of the passenger pigeon and the dodo (even with the latter's arguable lack of utility value) still saddens us.

These days, there is great debate over protecting jobs versus protecting endangered species. Perhaps some balance needs to be struck between the needs of humans and the needs of animals.

However, the Endangered Species Act will soon be up for re-authorization, and the administration would like to change it in a way that will be a loss for us all.

Currently, the Endangered Species Committee (known as the "God Committee" because it makes decisions that only the creator of the planet should be allowed to make) can be convened to override the Endangered Species Act–that is, to allow a development project to continue (for economic reasons, of course) even though its continuation may result in the destruction of a species of plant or animal from the face of the Earth. The administration wants to make it easier for the committee to conduct these overrides.

The EarthCards

U.S. Representatives: One card goes to your U.S. Representative, stressing the importance of his or her support for Senate Joint Resolution 206 and House Joint Resolution 418, and requesting that he or she fight any attempt towards exploitation of the Arctic National Wildlife Refuge. The card also supports strengthening, rather than weakening, the Endangered Species Act's re-authorization (thus restricting the "God Committee" as much as possible). If we had more room, we would have included one to each Senator. If you want, send a letter to each of your Senators, using our *EarthCard* as a model and substituting the words "Senate Joint Resolution 206" for "House Joint Resolution 418."

The Secretary of the Interior: The Interior Secretary should be the prime exponent of wilderness and wildlife in the nation. But current secretary Manuel Lujan's commitment, especially to the Endangered Species Act, appears uncertain. In a recent interview on the subject of an observatory proposed to be built in the habitat of the highly endangered Mount Graham Red Squirrel, he commented (rather after the fashion of ex-President Reagan's quip that "You see one redwood, you've seen them all") that perhaps it was not important to preserve red squirrels, as grey and black ones were quite plentiful. Our card to him will politely point out our disagreement on this subject, as well as requesting his support for the issues in the card just above.

Write for Action Resources: Antarctica: The Cousteau Society, 930 W. 21st St., Norfolk, VA 23517 (804) 627-1144. Arctic and Endangered Species Act: The National Wildlife Federation, 1412 16th St NW, Washington DC 20036, (202) 797-6800; World Wildlife Fund, 1250 24th St. NW, Washington, DC 20037 (202) 293-4800.

Energy Efficiency

The Problem: In its most general form, the problem is that as a nation, rather than conserve energy, we have traditionally preferred to seek additional resources to exploit, almost no matter what the financial, aesthetic, or moral cost. Thus many of us would rather risk the pristine eco-system of the Arctic or Antarctic to seek oil that may or may not be there, rather than reduce our consumption of energy so that our present supply is sufficient.

Naturally, when circumstances force us to conserve energy, we do–as we did after the "Gas Crisis of 1973" struck. From 1973 to 1986, the U.S. economy grew by 35%, while our energy consumption remained constant (instead of increasing by 35%, or 14,000,000 barrels of oil a day). So conservation doesn't mean "freezing in the dark," nor does it necessarily mandate lack of growth (although that might not be a bad idea). As energy expert Amory Lovins put it: "I'm not interested in doing with less. . . But in doing more with less. We don't need to become vegetarians and ride bicycles to save the Earth." (Although that might not be a bad idea either!)

The Solution: The answer lies, as Lovins said, in "doing more with less." A good place to start is gasoline consumption. Reducing our national craving for oil, much of which goes to fuel our automobile addiction, would improve national security by lessening dependence on foreign oil, reduce the budget deficit (Lovins estimates that one third of the U.S. military budget is directed towards maintaining access to foreign oil and minerals), and reduce pollution (autos are responsible for 20% of the nation's CO_2, a prime greenhouse gas, among other noxious effluents).

The EarthCards:

To the Automobile Manufacturers: Unfortunately, the average fuel efficiency of the nation's automobiles has been decreasing. Chrysler is discontinuing its Dodge Omni/ Plymouth Horizon line of small cars, and GM's Saturn (originally announced as a durable, affordable, 60 mpg sub-compact) has become a $10,000 plus, 35 mpg compact. On the other hand, Detroit seems increasingly eager to provide us with "muscle cars." The top speed offered by Ford's Mustang V8 (150 mph), GM's Corvette (160 mph) and Chrysler's upcoming Viper (180 mph) indicate that fuel efficiency and environmental impact were not high design priorities.

Ironically, technology currently in existence (like the multi-valve cylinders in use by most Japanese car makers, or the lean-burn engine) could easily provide mileage in excess of 45 mpg for most American cars. The American Council for an Energy Efficient Economy lists 17 improvements that could be inexpensively made to the transmission, engines, and aerodynamics of American cars. These could improve average new car fuel economy from today's 28 mpg to 46.5 mpg by the year 2000. Foreign manufacturers have even more ambitious plans–with prototypes that get from 77 to 107 mpg!

A non-mechanical method for reducing the impact of the auto on the earth is the "gas guzzler tax/gas sipper rebate" now being considered in several states. By penalizing the buyers of fuel-inefficient cars, and giving rebates to the buyers of efficient ones, this legislation supports both consumers and auto makers to make energy-wise decisions. Since the taxes and rebates will be calculated so as to equal each other, and the program instituted through sales tax boards already in existence, it will place no burden on taxpayers.

EarthCards will go to the CEOs of Ford, General Motors, and Chrysler, deploring their seemingly increased interest in production of high-speed "muscle cars" and expressing our consumer interest in high-mileage "eco-cars." We also request they support "gas guzzler tax/ gas sipper rebate" legislation on the national level.

Suggested P.S.: If you can truthfully say so, you might add something like: *Ecological considerations are a major component of my decision-making process when purchasing an automobile.*

To the Airlines: Detroit doesn't house the only adherents to the gas guzzler school of transportation. In 1987, U.S. airlines consumed 14.5 billion gallons of fossil fuel, up 38% from 1983's 10.5 billion gallons. By contrast, the growth in gasoline consumption by the entire U.S. automobile fleet was 9%, due in part to federal fuel economy standards.

There's a solution–but it's not being used. In 1988, aircraft maker McDonnell Douglas unveiled its new MD-91/92 plane, whose innovative design allowed it to consume roughly only 50% of the fuel required by typical planes of the same type and produce unburned hydrocarbon emissions that are 87% lower than the industry average, and carbon monoxide emissions that are 24% lower. We might have expected the airlines to express serious interest in the MD-91/92. But after a year of intensive marketing, not a single order had been placed and McDonnell Douglas was forced to cancel production.

EarthCards to the CEOs of two of the U.S.'s largest airlines, United and American, requests they consider purchasing MD-91/92s or similar low-impact aircraft when it comes time to replenish their fleets.

Write for Action Resources: NRDC: 40 West 20th St., NY, NY 10011 (212) 727-2700; National Wildlife Federation: 1412 16th St. NW, Washington, DC 20036 (202) 797-6800.

Corporate Commitment and Social Responsibility

The Problem: Some large corporations seem to have an interest in being environmentally responsible. Yet the realities of the American economic system are such that a publicly-held corporation is accountable to its stockholders for maximizing profit. However, certain socially conscious businesses, like Ben & Jerry's Ice Cream, have a dual bottom line accountability to their shareholders: to maximize profit while conducting business in a socially responsible manner.

For us, a simple explanation of what it means to be a "socially responsible" business would involve having a well thought out commitment to customers, employees, local community, nation and planet. Depth and expression would vary from business to business, but of course should include an environmental component. The highly reputable and broadly-based Coalition for Environmentally Responsible Economics (CERES) has provided such guidelines with its Valdez Principles. These are a series of ten environmental rules that we generally agree with: protecting the wilderness, dealing responsibly with toxic wastes, conserving energy, minimizing the greenhouse effect, acid rain, and depletion of the ozone layer, etc.

The Solution: If the bad news is that corporations are accountable to stockholders, that may also eventually prove to be the good news as well. With this idea in mind, we've chosen two corporations that seem to be concerned on some levels about the environment as recipients of a unique pair of *EarthCards*.

The EarthCards

To DuPont: E.I. Du Pont de Nemours & Company, although the nation's biggest releaser of toxic chemicals (at 338,416,705 pounds in 1988, according to *Consumer Action*), has recently announced its intention to reduce pollution by 50% by the year 2000; phase out ozone-destroying CFCs (by replacing them with somewhat better but not entirely benign HCFCs); and consider a public commitment to the Valdez Principles.

One card goes directly to Du Pont's CEO, regarding Du Pont's stated interest in adhering to the Valdez Principles and its commitment to reduction of toxic wastes, and alluding to the second card. The second card goes to the Du Pont Department of Stockholder Relations, to be communicated to shareholders. It will implore that they consider an environmental as well as a financial bottom line, so as to allow the corporation to make the changes discussed above.

To 3M: 3M Corporation won the 1988 environmental award from the Council on Economic Priorities, a public interest research group. It has a strong commitment to in-house recycling, as well as a vice president for environmental engineering who reports directly to the CEO each month (and who has publicly stated the importance of environmental considerations to 3M's shareholders, customers, and employees). However, 3M is also a major manufacturer of popular consumer items (like their Scotchgard® upholstery and fabric protector) containing the recently identified ozone-eater methyl chloroform. An *EarthCard* goes to the Chairman of the Board, expressing appreciation of 3M's interest in recycling, and requesting that it intensify research to identify methods of replacing methyl chloroform in its products. Another card will be sent to the Director of Investor Relations to be communicated to the shareholders, requesting that they consider an environmental as well as a financial bottom line.

Write For Action Resources: For more information on the Valdez Principles (including the names of co-operating businesses), please contact The CERES Project at 711 Atlantic Ave., Boston MA 02111 (617) 451-0927. If you'd like to learn more about business and social responsibility (as well as about investing money in a socially responsible way), contact the Social Investment Forum at 430 First Ave. #290, Minneapolis, MN 55401 (612) 333-8338.

Receiving an EarthCard Update

The Problem: How will you know whether *Earthcards* has had any effect? And how will we know what other versions of *EarthCards* you'd like us to produce?

The Solution: By sending in the last card, you can tell us what other *EarthCard* books you'd like, and sign up to receive a follow-up report on the effects of *EarthCards*. Your report will be produced by 20/20 Vision, a national organization that has agreed to monitor the effects of *EarthCards*. To receive a follow-up, send the card in an envelope along with a check or money order for $2 made out to 20/20 before 7/31/91. Your report will be mailed no later than 8/30/91.

Who is 20/20? 20/20 Vision is a national network of citizens who spend 20 minutes a month communicating to policy makers on environmental and peace issues. Members receive one postcard a month outlining a specific action they can take. Every six months, 20/20 follows up with a report describing the results of the action. 20/20 Vision is a perfect way to continue talking to decision makers about the issues you care about most. For more information, or to join the 20/20 Vision network, call (800) 347-2767. You can also join by sending a $20 annual membership fee in with the last *EarthCard*.

The Generic *EarthCards*

The Problem: Yes, sending *EarthCards* to politicians and CEOs is all very well–but what about the local issues that we encounter, but rarely find time to write about?

The Solution: Here are some generic *EarthCards* that can be used in a variety of situations. Simply photocopy them, cut the letters apart, paste them to a postcard or piece of stationary, sign and send. Send the recycling letter to your local city hall, addressed to the Director of Public Works; the media letter to local newspapers, and radio and tv stations, addressed to the editor or program director; the store letter to the manager; and the food vendor letter to the director of food services.

Dear Department of Public Works:
　　As a concerned consumer and resident of this community, I would like to be able to recycle aluminum, glass, paper, and, if possible, plastic in as convenient a way as possible. I'm writing to request that you implement/expand a curbside recycling program (or other appropriate recycling program) in our locality. I would also like to see a hazardous waste pickup or dropoff program begun, so that people can dispose of their toxics in a responsible manner. Thank you very much for your help in this matter.

Dear Local Media Programming Person:
　　As a local resident who has a strong concern for the environment, I'm writing to request that you offer more consistent environmental programming. It appears to me that with the exception of the week before Earth Day, the media pays little attention to the information needs of the many millions of watchers who are deeply concerned about ecological problems. Increased, ongoing, in-depth coverage of the major issues would make me turn to you for my information more often. I would also pay more attention to those who chose to advertise during such programming (as long as they were chosen with some concern as to environmental awareness). Thank you for your consideration on this matter.

Dear Merchant:

As a local consumer who has a strong concern for the environment, I want you to know that ecological issues have a powerful effect on my purchasing decisions. I prefer to shop in places that are sensitive to the environmental impact of packaging, and uses paper for wrapping rather than polystyrene. I am also favorably impressed when a store carries environmentally benign products.

Your purchasing decisions affect my purchasing decisions (and those of many other people who feel similarly). I hope that you will at least consider environmental issues in making your stocking and packaging choices. Thank you.

Dear Local Food Vendor:

Perhaps you haven't thought about it, but using washable dishes, cups, and silverware instead of disposable items can have an important effect on the environment. Most disposable foodware is made either of styrofoam (which not only is bad for the atmosphere, but is also completely unbiodegradable) or of plastic (which is nearly as bad). The environmental impact of creating and cleaning metal or ceramic foodware is typically less than that of making "single use" disposable items.

Please consider switching–many people like myself really prefer to eat without having to worry about the effect our eating implements will have on the environment! Thank you.

What To Do Next

Congratulations! By the time you've reached this... YOU HAVE MADE A DIFFERENCE! So what can you do next? Here are three suggestions:

1. Fill out and send the last *EarthCard*. It will let us know what other specific *EarthCard*-style books that you would be interested in so we can get started on putting them together, and will enable you to receive an update from 20/20 Vision on what effect *EarthCards* has had.

2. Call some of the organizations listed in the Write for Action Resources of each section, and get more involved in a particular local or national issue–they'll help you get started.

3. Have an *EarthCards* Party! If you buy six or more copies of EarthCards, we'll let you have each book for only $4 apiece plus a one-time charge of $2 shipping and handling! Buy a few rolls of stamps, invite a group of friends over, and have the kind of party that will make you feel better the next morning instead of worse: an *EarthCards* party!

Whether you'd like a single copy for a friend ($6.95 plus $2 shipping and handling, California residents please add $.46), enough *EarthCards* for a party (see just above), or information on bulk discount offers for groups and organizations, please contact:

The Write For Action Group
c/o Conari Press
1339 61st St.
Emeryville, CA 94608
(415) 596-4040

Dear President Bush:

When campaigning in 1988, you referred to yourself as the "Environmental President." It is imperative that you live up to that label, beginning today, because our nation is threatened by an enemy far more ominous than any foe we've faced to date. Poisoned air and water, global warming, acid rain, and the thinning of the ozone layer–these are our adversaries now. They must be countered with the same degree of human, financial, and technological commitment that America has mustered in times of war past. The waning of the Cold War provides you with immense resources that should be brought to bear upon the environmental emergency. Any lesser response will be ineffective.

Please understand that this issue is of tremendous personal importance to me, and will assuredly influence my future decisions at the ballot box.

Dear Congressperson:

As a constituent who is tremendously concerned about global warming, I am writing in support of H.R. 1078, the Global Warming Bill. If you already support or are a co-sponsor of this bill (as are 144 of your bi-partisan colleagues), please intensify your promotion as it goes through committee. If you do not currently support it, please re-consider your position. This is one of the most crucial issues facing the nation, and the time for action is past due.

Please do not underestimate the degree of my concern. Even though this is a pre-written postcard, consider it a reminder that I and thousands of others consider the global warming situation a top priority. Thank you for your sincere consideration of this vital issue.

_____
_____
_____

President George Bush
1600 Pennsylvania Avenue
Washington, DC 20500

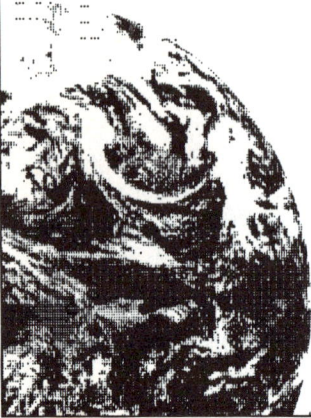

_____
_____
_____

The Honorable
House Office Building
Washington, DC 20515

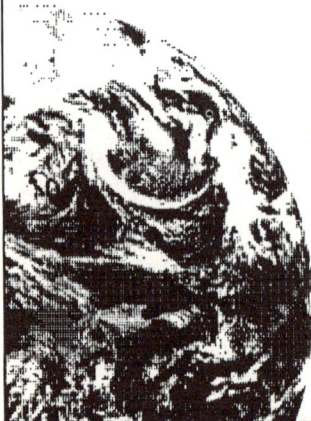

Dear Senator:

As a constituent who is tremendously concerned about global warming, I am writing in support of H.R. 1078, the Global Warming Bill. This is one of the most crucial issues facing the nation, and the time for action is long past due. I believe that a thorough study of the contents of H.R. 1078 will help create a single more comprehensive bill combining Senator Gore's S. 201 and Senator Wirth's S. 324 (thus removing emphasis on nuclear power, a costly and unnecessary part of the legislation).

Please do not underestimate the degree of my concern. Even though this a pre-written postcard, consider it a reminder that the global warming situation is so important to me that it cannot fail to influence my future behavior at the ballot box. I await your response, and am most interested in your position on this bill.

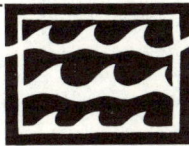

Dear Senator:

As a constituent who is tremendously concerned about global warming, I am writing in support of H.R. 1078, the Global Warming Bill. This is one of the most crucial issues facing the nation, and the time for action is long past due. I believe that a thorough study of the contents of H.R. 1078 will help create a single more comprehensive bill combining Senator Gore's S. 201 and Senator Wirth's S. 324 (thus removing emphasis on nuclear power, a costly and unnecessary part of the legislation).

Please do not underestimate the degree of my concern. Even though this a pre-written postcard, consider it a reminder that the global warming situation is so important to me that it cannot fail to influence my future behavior at the ballot box. I await your response, and am most interested in your position on this bill.

_____

_____

_____

The Honorable
Senate Office Building
Washington, DC 20510

_____

_____

_____

The Honorable
Senate Office Building
Washington, DC 20510

Dear Mr. Conable:

As a concerned American, I am writing to express my support for the World Bank's increasing consideration of environmental factors in the decision-making stage of its funding process. I would like to encourage you to expand upon your environmental committee, and to request that a "cradle-to-grave" analysis, with an emphasis on conservation of energy rather than generation of it, be applied to all potential projects before funding is considered.

I also believe that the World Bank should encourage "debt-for-nature" trades. Since our nation has a 17% stock ownership in the Bank, we must take a leadership role on this issue, which can both protect the global environment and relieve the pressing burden of debt on many Third World countries. Thank you for your attention to these urgent matters.

Dear Mr. Nicandros:

As an American very attentive to environmental issues, I am writing to thank you for sharing some of my concerns, as evidenced by your company's stated decision to double-hull your oil tankers. I would also like to emphatically request that Conoco reconsider its plan to build a road and oil pipeline into Ecuador's Yasuni National Park and the Waorani Indian Nation. Despite Conoco's intention to try to prevent damage to the native culture and the eco-system, the fact remains that whenever a road has been built into a rainforest, rampant and uncontrollable colonization follows.

A company's ecological track record will most certainly influence my behavior in the marketplace. I appreciate your company's actions on the first matter, and beg your re-consideration on the second.

Mr. Barber Conable
President
World Bank
1818 H. St. NW
Washington, DC 20433

Mr. Constantine Nicandros
President and C.E.O.
Conoco, Inc.
1007 Market Street
Wilmington, DE 19898

Dear Ms. Baer:

I would like to express my support of IHPA's attempts to be ecologically sensitive to the use of foreign hardwoods, and to request that IHPA reconsider its importation of hardwood from Burma and Malaysia. Most rainforest action groups agree that IHPA should refrain from purchasing hardwoods from this part of the world until the ecological and human rights issues have been better addressed, and the wood can be obtained by sustainable forestry practices. This includes wood from Thailand too, since Thailand acts as middleman for much of Burma's wood sales.

Please inform me of IHPA's actions on these requests, because rainforest protection is critical to me. Without assurances that these issues have been addressed, it will be difficult for me to buy any hardwood products.

Dear Congressperson:

I am writing regarding perhaps the most important upcoming piece of toxic waste legislation, the RCRA re-authorization. I truly believe that RCRA must emphasize *reduction* or *elimination* of hazardous waste, rather than attempting to regulate and minimize the hazards of disposal afterwards. The existing version seems to focus on building toxic waste incinerators, which many scientists believe will simply spread the poisons into the atmosphere, and on forcing states to create sites for hazardous waste dumping. A RCRA based on reduction will be better for the economy, because it will force companies to face the hidden costs of waste disposal that the government is now covering with our tax money.

Your plans regarding this legislation are of great interest to me, and I await your response. It will most assuredly affect my vote.

_____

_____

_____

Wendy Baer
Executive V.P.
Intl. Hardwood Products Ass.
P.O. Box 1308
Alexandria, VA 22313

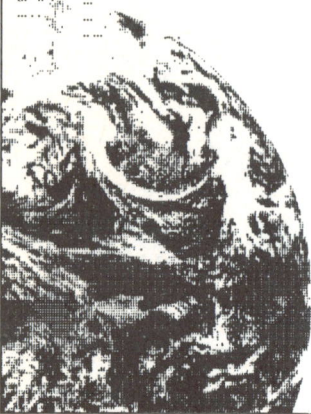

_____

_____

_____

The Honorable
House Office Building
Washington, DC 20515

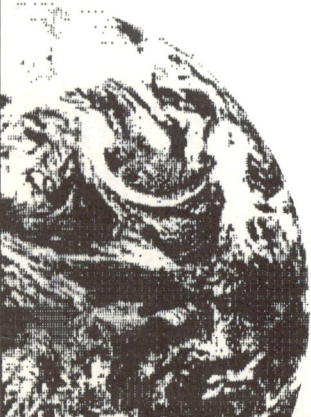

Dear Senator:

I am writing regarding perhaps the most important upcoming piece of toxic waste legislation, the RCRA re-authorization. I truly believe that RCRA must emphasize *reduction* or *elimination* of hazardous waste, rather than attempting to regulate and minimize the hazards of disposal afterwards. The existing version seems to focus on building toxic waste incinerators, which many scientists believe will simply spread the poisons into the atmosphere, and on forcing states to create sites for hazardous waste dumping.

I plan to take an active interest in this legislation, so your stand is very important to me. I will then be able to better inform my family, friends, and neighbors so that they can make choices in the voting booth that reflect our concern with the health of the nation's air, food, and water.

Dear Senator:

I am writing regarding perhaps the most important upcoming piece of toxic waste legislation, the RCRA re-authorization. I truly believe that RCRA must emphasize *reduction* or *elimination* of hazardous waste, rather than attempting to regulate and minimize the hazards of disposal afterwards. The existing version seems to focus on building toxic waste incinerators, which many scientists believe will simply spread the poisons into the atmosphere, and on forcing states to create sites for hazardous waste dumping.

I plan to take an active interest in this legislation, so your stand is very important to me. I will then be able to better inform my family, friends, and neighbors so that they can make choices in the voting booth that reflect our concern with the health of the nation's air, food, and water.

_____
_____
_____

Postcard
Stamp
Here

The Honorable
Senate Office Building
Washington, DC   20510

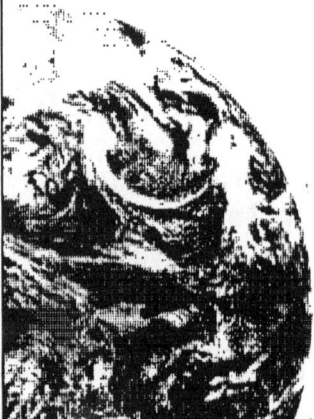

_____
_____
_____

Postcard
Stamp
Here

The Honorable
Senate Office Building
Washington, DC   20510

Dear Mr. Kidder:

As an environmently-conscious consumer, I am writing to express my appreciation of the household battery industry's dramatic reduction in the use of mercury since 1983. I am also asking you to escalate your commitment to the environment by beginning to use European mercury-free battery technology in your manufacturing process.

Additionally, Duracell should implement a disposable alkaline battery recycling program by offering to buy back used batteries (which when available in sufficient numbers could then be economically recycled). Such a program could be modeled on the current K-Mart program for recycling car batteries and economically set up through one of your national retailers.

My ecological concerns have a strong effect on my behavior as a consumer, and I would far prefer to buy the brand of batteries that attempts to implement the above suggestions.

Dear Mr. Mulcahy:

As an environmently-conscious consumer, I am writing to express my appreciation of the household battery industry's dramatic reduction in the use of mercury since 1983. I am also asking you to escalate your commitment to the environment by beginning to use European mercury-free battery technology in your manufacturing process.

Additionally, Eveready should implement a disposable alkaline battery recycling program by offering to buy back used batteries (which when available in sufficient numbers could then be economically recycled). Such a program could be modeled on the current K-Mart program for recycling car batteries and economically set up through one of your national retailers.

My ecological concerns have a strong effect on my behavior as a consumer, and I would far prefer to buy the brand of batteries that attempts to implement the above suggestions.

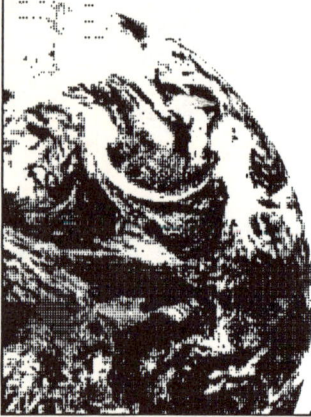

Postcard
Stamp
Here

Mr. C. Robert Kidder
President and C.E.O.
Duracell, Inc.
Berkshire Industrial Park
Bethel, CT   06801

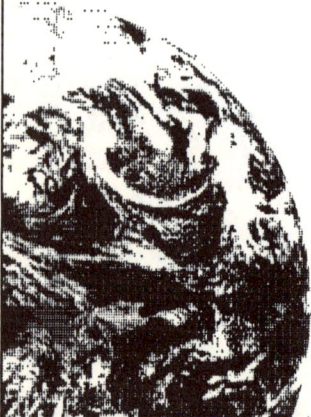

Postcard
Stamp
Here

Mr. J.P. Mulcahy
Chairman of the Board and C.E.O.
Eveready Battery Company
Checkerboard Square
St. Louis, MO 63164

Dear Mr. Hoeft:

As a concerned book buyer, I'd like to express my support for your company's stated interest in the use of recycled paper for some books and encourage you to do this as much as possible. I believe millions of Americans like myself would prefer to buy books made with the highest possible percentage of post-consumer recycled waste paper, even if that meant a slight reduction in paper quality.

Were BDD to use a 50% recycled stock for even half of the books published each year, it would result in a savings of an estimated 250,000 trees and 100,000,000 gallons of water annually, and eliminate 850,000 pounds of air polluting effluent and 43,000 cubic yards of solid waste. By using other low-impact publishing techniques such as soy ink and recyclable binding glue, BDD could have an even greater positive effect on the environment.

Dear Mr. Yastrow:

I appreciate your company's new McRecycle USA program and your willingness to work with environmental organizations, especially given the size and visibility of McDonald's. But I feel there are other steps you should take. First, McDonald's needs to reduce its total amount of packaging per food item sold. You must also reconsider the decision to begin installing incinerators, since surveys show that incineration of styrofoam (especially when mixed with other food and packaging materials) can release a variety of toxic chemicals.

Lastly, McDonald's needs to offer a non-meat burger, a "McVeggieBurger," and thus lead the nation's fast food chains in offering an ecological alternative to beef (production of vegetables has much less impact on the earth's limited resources than beef). Please show the sincerity of your environmental stance by actions on these suggestions.

Postcard
Stamp
Here

Mr. Jack Hoeft
President
Bantam/Doubleday/Dell
666 Fifth Ave, 25th Floor
NY, NY   10103

Postcard
Stamp
Here

Shelby Yastrow
V. P. for Environmental Affairs
McDonald's Corporation
1 McDonald's Plaza
Oakbrook, IL 60521

Dear Mr. Smale:

I'm writing to express encouragement for P&G's increased interest in the environment, and request that you use the 100% recycled plastic in the new NorthWest market's Spic and Span bottles as much as possible in other packaging. I'd also like to suggest some products that I (and millions of other environmentally-aware consumers) would like to see on every supermarket shelf in the nation. These are: a non-phosphate powdered laundry detergent; a cotton-based diaper insert for use with reusable diaper covers; a line of vegetable-based cleaning agents for dishwashers, windows, woolens, floors, and cars; and a line of high post-consumer waste content (100% if possible), unbleached paper products including toilet paper, facial tissue, and paper towels. These products will help the environment, and your sales!

Dear Mr. Kleckner:

As an environmentally-concerned consumer, I read with great interest of the Farm Bureau's recent poll in which three out of four Americans expressed concern about the amount of pesticides used by farmers. I read with even greater interest, and dismay, your response.

We don't need to be "educated" on pesticide safety. We know that excessive pesticide use is unsafe, and that farmers have dramatically increased pesticide use since World War II, with equally dramatic decreases in pesticide effectiveness. Low-impact farming techniques such as integrated pest management have proven their worth and the public has shown its interest in organic and pesticide-free produce. The mandate of the Farm Bureau is to help the American farmer. You can best do that by providing your members with the motivation and information they need to provide safe and healthy food. We want pesticide-free produce, not a public relations campaign!

_____

_____

_____

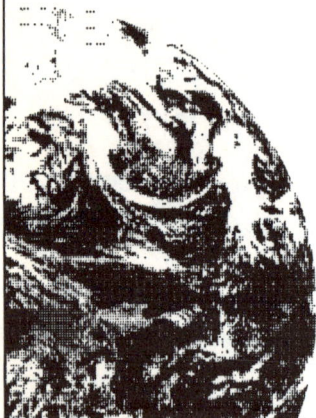

Mr. J.G. Smale
Chairman of the Board and C.E.O.
The Proctor & Gamble Company
One Proctor & Gamble Plaza
Cincinnati, OH 45202

_____

_____

_____

Dean R. Kleckner
President
American Farm Bureau Fed.
225 Touhy Ave.
Park Ridge, IL 60068

Dear Mr. Lippincott:

As an ecologically-concerned consumer, I'd like to suggest some products that I (and many millions of other environmentally-aware consumers) would like to see on every supermarket shelf in the nation. These include a line of high (100% if possible) post-consumer waste content, unbleached paper products such as toilet paper, facial tissue, paper towels, and picnic ware, as well as a cotton-based diaper insert for use with reusable diaper covers. I also believe that a cup containing no polystyrene that could be used for hot and cold liquids would be tremendously popular. Such products, manufactured with sincere attention to their environmental impact, would be tremendously attractive to me, and strongly influence my purchasing decisions.

Dear Mr. Tisch:

As an American concerned with the environment, I'm writing to request that CBS offer more consistent environmental programming. It appears to me that, with the exception of the week before Earth Day, CBS pays little attention to the information needs of the many millions of viewers who are deeply concerned about the environment. When a particularly gruesome disaster happens (like the Exxon Valdez oil spill), it becomes a brief media event. But continuing catastrophes like the worldwide destruction of the rainforest or global warming are largely ignored, except for an occasional special.

Increased, ongoing, in-depth coverage of major issues would not only dispose me to turn to CBS more often, but would also incline me to look favorably on the sponsors (as long as they were chosen with some concern as to environmental awareness).

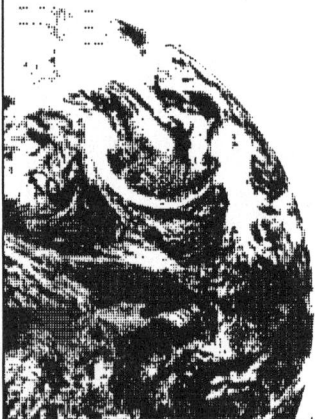

_____
_____

Postcard
Stamp
Here

Mr. Philip E. Lippincott
President and C.E.O.
Scott Paper Company
Scott Plaza
Philadelphia, PA 19113

_____
_____

Postcard
Stamp
Here

Mr. Lawrence Tisch
President
CBS
51 West 52nd Street
NY, NY 10019

Dear Mr. Tartikoff:

As an environmentally-concerned American, I'm writing to request that NBC offer more consistent environmental programming. It appears to me that, with the exception of the week before Earth Day, NBC pays little attention to the information needs of the many millions of viewers who are deeply concerned about the environment. When a particularly gruesome disaster happens (like the Exxon Valdez oil spill), it becomes a brief media event. But continuing catastrophes like the worldwide destruction of the rainforest or global warming are largely ignored, except for an occasional special.

Increased, ongoing, in-depth coverage of major issues would not only dispose me to turn to NBC more often, but would also incline me to look favorably on the sponsors (as long as they were chosen with some concern as to environmental awareness).

Dear Mr. Fias:

As an environmentally-concerned American, I'm writing to request that ABC offer more consistent environmental programming. It appears to me that, with the exception of the week before Earth Day, ABC pays little attention to the information needs of the many millions of viewers who are deeply concerned about the environment. When a particularly gruesome disaster happens (like the Exxon Valdez oil spill), it becomes a brief media event. But continuing catastrophes like the worldwide destruction of the rainforest or global warming are largely ignored, except for an occasional special.

Increased, ongoing, in-depth coverage of major issues would not only dispose me to turn to ABC more often, but would also incline me to look favorably on the sponsors (as long as they were chosen with some concern as to environmental awareness).

_____

_____

_____

Postcard
Stamp
Here

Mr. Brandon Tartikoff
President
NBC
3000 West Alameda
Burbank, CA 91523

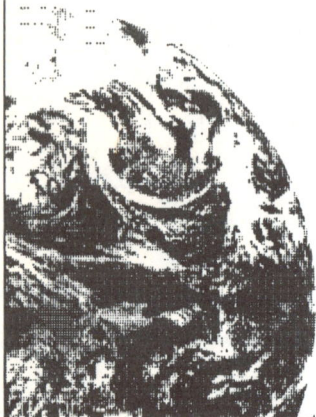

_____

_____

_____

Postcard
Stamp
Here

Mr. John Fias
President
ABC
2040 Avenue of the Stars
Los Angeles, CA 90067

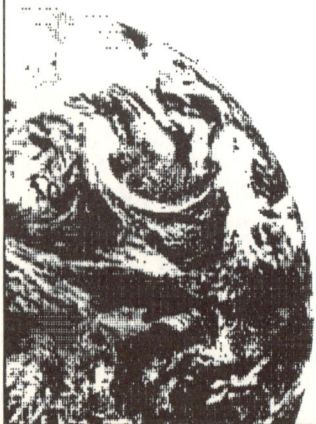

Dear Congressperson:

I am one of your many constituents who is extremely concerned about the preservation of wilderness areas. I would like to request your support for House Joint Resolution 418–which would reinforce world efforts towards making the Antarctic an international park–exactly what it should be. I further ask that you fight any attempt to allow oil or mineral exploitation of the Arctic National Wildlife Refuge, or any weakening of the Endangered Species Act.

I believe energy conservation efforts can obviate any need to exploit wilderness areas. Maintaining these areas in as near a pristine state as possible, and preserving species diversity as well, is of tremendous importance to me. Knowing where you stand on these issues will help determine how I vote.

Dear Mr. Lujan:

I feel it is crucial that, as Secretary of the Interior, you honor your mandate as protector of our nation's precious natural resources. These include far more than the oil, trees, and minerals that can be taken from the Earth. Even more valuable, and infinitely harder to replace, are America's wilderness areas and diversity of animal and plant species.

Millions of concerned citizens like myself look to you to make the difficult decisions that balance endangered species against jobs, or wildlife refuge areas against oil exploration and lumber exploitation. But it's not only we who judge your decisions–the historians of the future will judge as well, so your choices must reflect a long-term perspective. Thus we beg you to support strengthening of the Endangered Species Act rather than the "God Committee." It is not the place of men or women to decide which creatures survive, and which will not. Please also lend your support to HJR 418 and SJR 206.

```
_____
_____
_____
```

Postcard
Stamp
Here

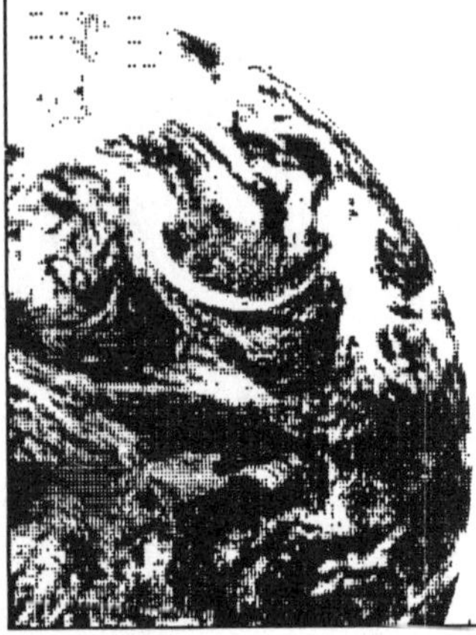

The Honorable
House Office Building
Washington, DC   20515

```
_____
_____
_____
```

Postcard
Stamp
Here

Mr. Manuel Lujan
Dept. of the Interior
1800 C Street
NW Washington, DC 20240

Dear Mr. Poling:

I am a potential Ford car buyer with a strong interest in environmental issues. I'm very concerned with what I see as a tendency to produce high speed "muscle cars" (like your 150 mph Mustang V8) when what America really needs is a 80 mpg "eco-car"! Even with current technology (like the multi-valve cylinder or the lean-burn engine), most American cars could easily get close to 50 mpg with sufficient power for all reasonable uses (foreign prototypes already double that figure). And innovations like the variable displacement and the stratified charge engine, and the electric car that could maximize efficiency and minimize pollution appear to be within our technological grasp. We also urge you to support non-mechanical methods for increasing fuel efficiency, like the gas guzzler tax/gas sipper rebate.

I, like many millions of consumers, would like a safe, economical, fuel-efficient American car–and hope that Ford can meet these needs in the very near future.

Dear Mr. Stempel:

I am a potential GM car buyer with a strong interest in environmental issues. I'm very concerned with what I see as a tendency to produce high speed "muscle cars" (like your 160 mph Corvette) when what America really needs is a 80 mpg "eco-car"! Even with current technology (like the multi-valve cylinder or the lean-burn engine), most American cars could easily get close to 50 mpg with sufficient power for all reasonable uses (foreign prototypes already double that figure). And innovations like the variable displacement and the stratified charge engine, and the electric car that could maximize efficiency and minimize pollution appear to be within our technological grasp. I also urge you to support non-mechanical methods for increasing fuel efficiency, like the gas guzzler tax/gas sipper rebate.

I, like many millions of consumers, would like a safe, economical, fuel-efficient American car--and hope that GM can meet these needs in the very near future.

Mr. H.A. Poling
Chairman of the Board and C.E.O.
Ford Motor Company
The American Road
Dearborn, MI 48121

Mr. Robert C. Stempel
President and C.E.O.
General Motors
3044 W. General Motors
Dearborn, MI 48202

Dear Mr. Iacocca:

I am a potential Chrysler car buyer with a strong interest in environmental issues. I'm very concerned with what I see as a tendency to produce high speed "muscle cars" (like your 180 mph Viper) when what America really needs is a 80 mpg "eco-car"! Even with current technology (like the multi-valve cylinder or the lean-burn engine), most American cars could easily get close to 50 mpg with sufficient power for all reasonable uses (foreign prototypes already double that figure). And innovations like the variable displacement and the stratified charge engine, and the electric car that could maximize efficiency and minimize pollution appear to be within our technological grasp. I also urge you to support non-mechanical methods for increasing fuel efficiency, like the gas guzzler tax/gas sipper rebate.

I, like many millions of consumers, would like a safe, economical, fuel-efficient American car–and hope that Chrysler can meet these needs in the very near future.

Dear Mr. Wolf:

I have recently become aware that in 1988, McDonnell Douglas unveiled its MD-91/92, whose innovative design allowed it to consume roughly 50% of the fuel required by planes of the same type. With emissions of unburned hydrocarbons 87% lower than the industry average and carbon monoxide emissions 24% lower, I would have expected United Airlines' enthusiastic interest in this craft. After all, your own inflight magazine paid homage to Earth Day 1990, and the MD-91/92 sounds like a great opportunity to reduce both fuel expenditures and pollution in these days of uncertain fossil fuel costs and impending Clean Air Act action.

I am most interested in your response, as the environment is of great concern to me, and an airline whose fleet was designed to minimize environmental impact would attract my consumer dollars over its less eco-conscious competition.

Mr. Lee Iacocca
Chairman of the Board and C.E.O.
Chrysler Motors
12000 Chrysler Drive
Dearborn, MI 48288

Stephen M. Wolf
CEO and Chairman of the Board
United Airlines, Inc.
1200 Algonquin
Elk Grove Village, IL 60007

Dear Mr. Crandall,

I have recently become aware that in 1988, McDonnell Douglas unveiled its MD-91/92, whose innovative design allowed it to consume roughly 50% of the fuel required by planes of the same type. With emissions of unburned hydrocarbons 87% lower than the industry average and carbon monoxide emissions 24% lower, I would have expected American Airlines' enthusiastic interest in this craft. After all, your own inflight magazine paid homage to Earth Day 1990, and the MD-91/92 sounds like a great opportunity to reduce both fuel expenditures and pollution in these days of uncertain fossil fuel costs and impending Clean Air Act action.

I am most interested in your response, as the environment is of great concern to me, and an airline whose fleet was designed to minimize environmental impact would attract my consumer dollars over its less eco-conscious competition.

Dear Mr. Wollard:

As an environmentally-concerned consumer, I appreciate Du Pont's commitment to reduce pollution by 50% by the year 2000, to phase out CFCs (even though the replacement HCFCs are not entirely benign), and to consider a commitment to the Valdez Principles, which I urge you to adopt.

I fully understand that the realities of the American economic system are such that a publicly-held corporation is accountable to its stockholders for maximizing profit. However, it is possible to maximize profit while conducting business in a socially responsible manner. An *EarthCard* similar to this one is being sent to your Department of Stockholder Relations, which may help influence your investors to support your worthy goals.

Robert L. Crandall
CEO and Chairman of the Board
American Airlines, Inc.
4200 American Blvd.
Forth Worth, TX 76155

Mr. Edgar J. Wollard
C.E.O.
Du Pont Company
1007 Market Street
Wilmington, DE 19898

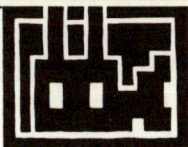

Dear Department of
Stockholder Relations:

I appreciate Du Pont's recent commitment to reduce pollution by 50% by the year 2000, phase out CFCs (even though the replacement HCFCs are not entirely benign), and consider a public commitment to the Valdez Principles (which I urge you to adopt).

Since I understand that a publicly-held corporation is accountable to its stockholders for maximizing profit, I am actually writing this card to Du Pont's stockholders, and ask that you pass my message on to them: I urge that you consider a dual bottom line accountability that involves maximizing profit while conducting business in a socially responsible manner (for an example of this, please see ice cream manufacturer Ben & Jerry's annual report). By supporting Du Pont in its movement towards environmental accountability, you are in the forefront of the biggest battle that the planet has ever faced. Thank you for considering the survival of us all as part of your "bottom line."

Dear Mr. Jacobson:

As a ecologically-concerned consumer, I appreciate 3M's strong commitment to in-house recycling, as well as V.P. Robert Bringer's stated commitment to environmental considerations for 3M's shareholders, customers, and employees. However, I would also like to see 3M intensify research to replace methyl chloroform in your products, and consider adopting the Valdez Principles.

I fully understand that a publicly held corporation is accountable to its stockholders for maximizing profit. However, it is possible to maximize profit while conducting business in a socially responsible manner. An *EarthCard* similar to this one is being sent to your Director of Investor Relations, which may help influence your investors to support your environmental goals.

Postcard
Stamp
Here

Du Pont Stockholders
C/O Du Pont Company
Dept. of Stockholder Relations
1007 Market Street
Wilmington, DE 19898

Postcard
Stamp
Here

Mr. Allen Jacobson
Chairman of the Board and C.E.O.
3M Company
3M Center 220-14W-04
St. Paul, MN   55144

Dear Ms. McBride:

As a environmentally-concerned consumer, I appreciate 3M's ecological commitment. However, greater steps need to be taken by 3M to protect the environment, perhaps the most important of which is to come up with safe replacements for ozone destroying methyl chloroform in your products, and to consider adopting the Valdez Principles.

Since I understand that a publicly held corporation is accountable to its stockholders for maximizing profit, I ask that you pass this message on to 3M investors: Please consider a dual bottom line accountability that involves maximizing profit while conducting business in a socially responsible manner (for an example, please see ice cream manufacturer Ben & Jerry's annual report). By supporting 3M in its movement towards environmental accountability, shareholders are in the forefront of the biggest battle that the planet has ever faced. Thank you for considering the survival of us all as part of your "bottom line."

Dear Write for Action:
I would be interested in EarthCard books on the following:

____Global Warming/Ozone Deterioration ____Rainforest Preservation
____Toxic Waste ____Recycling
____Ecological Consumer Products ____Wilderness Preservation
____Corporate Commitment and ____Energy Efficiency
    Social Responsibility

_____Please sign me up to receive a special follow-up report on the effects of EarthCards. I've enclosed a check for $2 to help defray costs made out to 20/20 Vision. I realize that my report will not be compiled until summer 1991.

Name_____

Address_____

       _____

Send this card to: 20/20 Vision, 69 South Pleasant St. #203, Amher

Ms. Joan McBride
Director of Investor Relations
3M Company
3M Center
225-5N-04
St. Paul, MN 55144

Please send this card in an envelope to the
address on the reverse! Thanks!!!